U.S.NRC

United States Nuclear Regulatory Commission

Protecting People and the Environment

NUREG-1954

I0475786

Enterprise Content Management

Streamlining How Staff and Stakeholders Work within the Nuclear Regulatory Commission's Regulatory Environment

Manuscript Completed: September 2010
Date Published: September 2010

Office of Information Services

ABSTRACT

This report provides staff and stakeholders with an understanding of Enterprise Content Management (ECM), describes the U.S. Nuclear Regulatory Commission's (NRC's) planned ECM Program, and discusses how this ECM Program will provide critical support to the NRC's mission and streamline the work of staff and stakeholders within the NRC's regulatory environment. ECM is the ability of an organization to capture, manage, store, preserve, and deliver needed information so that it is available to the right people at the right time.

CONTENTS

EXECUTIVE SUMMARY

The primary mission of the U.S. Nuclear Regulatory Commission (NRC) is to regulate civilian nuclear energy activities to protect people and the environment and promote the common defense and security. As an independent regulatory agency, the NRC is responsible for determining if nuclear material is being used consistently with Commission regulations. The NRC licenses individuals and organizations to use radioactive materials. There are 104 commercial nuclear power reactors, three-quarters of which either have received or are under review for license renewal; 32 research and test reactors; approximately 4,600 licensed reactor operators; 4 early site permits for potential new reactors issued, 2 applications, and 2 more expected by 2012; 4 reactor design certifications issued and 5 under review; 18 combined construction and operating license applications for 28 new reactor units, with additional applications expected over the next several years; 5 uranium recovery sites and approximately 15 applications for new sites expected by 2013; 15 major fuel cycle facilities; approximately 3,000 research, medical, industrial, government, and academic materials licenses; 55 independent spent fuel storage installations; numerous sites undergoing decommissioning and numerous transportation related activities each year; and an application from the U.S. Department of Energy (DOE) for a high-level waste repository. (Of the 18 applications for combined construction and operating licenses, five applicants have requested that the NRC suspend its review.)

Information is an important agency asset, second only to the NRC staff. The staff carries out its regulatory programs through the ongoing receipt, development, analysis, and documentation of information. Thus, timely and secure access to and use of reliable information is critical to the NRC staff's ability to provide efficient and effective regulation of the above facilities, as well as agency support operations. It is a vital part of an effective and efficient regulatory program, that at any time or place, the staff is able to have immediate and secure access to information and the tools to use information related to a licensed activity, site, or radioactive material. This may include access to information on similar activities, sites, and material.

To improve the management and use of the vast amount of information used by the agency, the NRC is undertaking an Enterprise Content Management (ECM) Program. The ECM Program serves as a trusted framework to help the agency achieve its regulatory mission and to empower the staff to use information more easily, as needed, in different functional areas. Readily available and reliably provided information directly supports the NRC safety and security goals.

In particular, ensuring the availability of secure information helps the NRC to be a stable, credible, and predictable regulator. Giving the NRC staff and stakeholders access to needed information creates the foundation for effective and structured regulatory processes. The ability to research and identify prior NRC actions and precedents, find information on generic safety issues related to a particular piece of equipment or material use, or locate staff guidance documents showing how compliance with NRC regulations can be achieved, makes it possible for the NRC staff and stakeholders to have the information needed so that new regulatory and licensing actions are fully informed by all relevant information.

Ready and secure access to complete and current information directly supports the NRC safety and security goals as specified in NUREG-1614, Volume 4, "The Strategic Plan: Fiscal Years 2008–2013," issued February 2008, leading to stable and predictable regulatory policies and programs. Predictable regulatory programs and policies support the following strategies in the safety goal: (1) develop, maintain, and implement licensing and regulatory programs…to ensure adequate protection of public health and safety and the environment (Strategy 1), (2) continue to oversee the safe operation of existing plants, while preparing for and managing the review of applications for new power reactors (Strategy 2), and (3) promote focused attention on safety matters (Strategy 6).

This NUREG provides insights into the NRC ECM Program. It shows how ECM is an integral component in achieving the safety and security strategies and excellence objectives contained in the NRC's Strategic Plan. In addition, the report shows how ECM supports and empowers the NRC staff in performing its responsibilities to protect the public and the environment and promote the common defense and security. It shows how staff will save time searching for relevant material, how the staff can create automated work flows and virtual workspaces and use collaboration tools like SharePoint, and how the staff will be able to work from anywhere.

Specific ECM projects tie directly to staff activities such as license reviews and inspection. Also, ECM will address timely communication and coordination of information with State, local, and Federal (DOE and Department of Homeland Security) agencies, Agreement States, and international counterparts. Finally, ECM gives NRC stakeholders timely access to information so that they can more effectively participate in the agency's regulatory and licensing programs. As discussed in this NUREG, some of the components of, and some of the activities that support, the ECM Program are currently being implemented; implementation of others will depend on availability of budgeted resources.

ABBREVIATIONS

ADAMS	Agencywide Documents Access and Management System
DOE	U.S. Department of Energy
ECM	Enterprise Content Management
FOIA	Freedom of Information Act
IM	information management
IT	information technology
NRC	U.S. Nuclear Regulatory Commission
OIS	Office of Information Services
OCFO	Office of the Chief Financial Officer

1. INTRODUCTION

1.1   Effective Enterprise Content Management

Agency information takes many forms and resides in many places. Today, much agency information is generated electronically and may be either structured (data in spreadsheets or databases) or unstructured (e-mail files, text documents, media files). It may be in the form of design specifications and drawings, hearing transcripts, scientific datasets, microfiche cards, or community-of-practice discussion threads. Agency information lives today in the Agencywide Documents Access and Management System (ADAMS); on network drives; on the agency's public and internal Web sites; in collaborative environments, like SharePoint; in specific information technology (IT) systems, such as the National Source Tracking System; and in boxes of paper records stored by the National Archives and Records Administration.

The current challenge for the U.S. Nuclear Regulatory Commission (NRC) is to smartly manage and enable the use of all this content or information from an enterprisewide perspective. The NRC staff relies on timely, secure, and reliable access to information to support its many activities on both the mission and corporate support side. In mission offices and regions, staff must have information to be able to conduct licensing reviews, perform inspections, support agency hearings, and respond to incidents and events. In corporate support organizations, staff must have information to put in place contracts, process financial invoices, develop and support the agency budget, and conduct many and varied administrative activities.

Enterprise Content Management (ECM) is a program to take control of the NRC's information and facilitate its move to a more automated environment. This shift is revolutionary and necessary to do the NRC's work, given the new licensing activities facing the agency. The NRC is focusing its ECM Program development on the following:

- empowering staff by providing flexible tools so that staff members can determine how to best carry out and streamline their activities and business processes

- improving staff productivity by managing content in a way that ensures its availability to help achieve mission or corporate support activities

- ensuring the NRC's ability to manage content in a way that integrates records, including automated records management, and information management into the processes for conducting the NRC's regulatory and licensing work

- providing a "reduced sign-on" and single intuitive interface for all content-related material

- providing a "single search" capability for electronically stored information

- defining policies and processes to ensure that content is managed throughout its life cycle and that the integrity and confidentiality of information is maintained

- reducing the agency's risk associated with continued use of aging technologies

- implementing new processes and technologies to address several current challenges and organizational drivers

ECM is a trusted program for managing Agency information across the enterprise. ECM embeds every system, tool, and automated or manual process that deals with any structured or unstructured content. The main goal of the ECM Program is to provide the technologies, methodologies, tools, and business processes to manage information (capture, utilize, store, preserve, and find) in a way that aligns with the business needs and regulatory responsibilities of the NRC's program and support offices. It is expected that, once fully implemented, the ECM Program will offer seamless access to electronic information, regardless of where it is stored. The ECM framework is a functional structure and operates independently of the NRC organizational structure. However, it is integrated with NRC mission and corporate support programs to ensure that ECM is a component of agency operations and not merely a support function.

The features of the NRC's ECM Program will empower staff to improve its content-oriented business processes through workflow development and/or modification, the use of collaboration software, and creation of virtual workspaces. Implementation of these improvements to date has been limited and very time consuming. The workflows currently in use at the agency automate manual tasks and activities, thereby increasing staff efficiency and effectiveness. Where there are currently no workflows, staff may not fully realize where bottlenecks and barriers exist. With the features provided in ECM, the staff will be able to implement these improvements on its own. With the NRC's ECM Program, NRC staff and stakeholders will also be able to better understand what content is available and why standard metadata (data that describe information, such as author and title), and taxonomies (a categorization framework that organizes content) are used for classification of content. The functionality provided in the NRC's ECM Program will also facilitate stakeholder participation in the NRC's regulatory processes. The ECM Program that the NRC is developing will also eventually allow for automated records management and a "single search" by both staff and stakeholders for all content in the agency's various content repositories. ECM can be fully successful only with collaboration among the Office of Information Services (OIS) and the programmatic and administrative support organizations, including the regions, and with coordination and integration of activities by all NRC organizational units involved in a particular programmatic or administrative support functional area.

1.2   How ECM Will Help Staff and Stakeholders

The ECM Program will help the NRC establish strong, independent, predictable, and stable regulatory policies and programs, as called for in the NRC Strategic Plan. The program also helps ensure that staff can effectively and efficiently coordinate its activities with coworkers, counterparts outside the agency, and stakeholders. The more efficiently and effectively the staff conducts its work, the sooner that work will become available to stakeholders. The complex and changing environment associated with the use of nuclear materials in the United States requires more effective and open communication, not only with the NRC's stakeholders but also within the agency. Because reliable and timely access to information is a cornerstone of meaningful stakeholder involvement, sound and secure management of information content helps to ensure confidence in the regulatory process. With its goals of improved searching and workflow, increased accessibility via the Web, streamlined profiling and classification of documents, increased consistency of information, and improved management of information throughout its life cycle, the ECM Program plays a significant role in helping the agency meet current and future challenges and in allowing the staff to spend more time on technical and programmatic activities.

The ECM Program will help achieve the "means of support safety and security strategies" in the NRC Strategic Plan to establish and maintain stable and predictable regulatory programs and policies. By moving the NRC from paper-based processing of information to electronic processing of information, ECM will ensure that information can be recovered when wanted by staff or stakeholders, that staff can work from anywhere, and that automated business processes and workflows can be established to most effectively and efficiently use and share this information. This will make possible scenarios that include the following:

- Even if he is on the road visiting a construction site, a mechanical engineer can be confident that he is viewing on his laptop the latest version of a chapter on instrumentation and controls for the AP1000 in support of a combined license application,

- An auditor will be able to search for data relevant to a case and be assured of finding all of the information, no matter where it resides, with greater ease and speed.

- Staff members from different offices who are working on a particular activity or project will have the flexibility to create workflows that best support their needs and virtual and/or collaborative workspaces that will allow them to conduct their work more efficiently and effectively.

- More efficient work by staff will, in turn, allow for more timely access to information by stakeholders.

- By conducting a single search of the system, a Freedom of Information Act (FOIA) coordinator answering a FOIA request can be assured of retrieving all the documents relevant to the request and doing it more efficiently.

- Similarly, the use of a single-search capability will give stakeholders confidence that they are gaining access to all relevant publicly available documents. This should limit, to some extent, the scope and number of FOIA requests.

- An engineer who needs a technical specification for a power plant built in 1974 is frustrated until she is informed that another engineer who worked on the original review has a copy in his office. In the future, older agency documents that are difficult to locate will be easier to find because of the digitization and indexing efforts that are part of the ECM Program.

- A resident inspector who adds an e-mail to ADAMS will not have to worry about whether the record properly reflects the correct docket number or the correct author or how long the e-mail must be retained. Automated e-mail capture can by design accomplish these tasks, enabling stable and predictable indexing of these important documents, which will give the staff more time to concentrate on mission-critical activities.

- As NRC staff has already demonstrated, SharePoint's collaborative features can be used to collect information on the current and future budget calls. Under the ECM

Program, content in SharePoint sites will be incorporated into enterprisewide content management processes.

- A nuclear engineer can confidently work on her draft regulatory issue summary document from home, knowing that her colleagues can access the most current version as soon as she saves it to the team's SharePoint site. These colleagues can then comment on her draft and save their comments on the SharePoint site for her to see.

- Automated e-mail capture of public comments regarding environmental impact statements will enrich stakeholder involvement, helping to ensure public confidence that the environment is adequately protected.

- While in the field, an inspector can determine whether something he has identified at a particular type of facility has been identified at similar facilities. He can also access the operational data and history of the facility being inspected and that of similar facilities.

- A stakeholder will be able to more easily participate in the NRC's regulatory processes, especially when single search is implemented. A stakeholder searching for information relevant to an activity, project, or licensee can be assured of finding all of the information, no matter where it resides, with greater ease and faster turnaround.

- A license reviewer can access seismic hazard analyses performed for the same or similar seismic areas and identify known issues and whether they have been resolved.

- A staff member in one agency organization can have access to work other staff members in a different organization(s) are performing on the same or similar projects, through collaborative work environments, enhanced search, and better records management.

- A license reviewer can be assured that, through single search, she is aware of all relevant information related to a proposed licensing action that the NRC possesses, including relevant information from similar licensing actions.

- Staff members, whether an inspector preparing to go into the field, a technical reviewer searching for information relevant to the review of a license amendment request, or a budget analyst evaluating trends over a number of years, will spend considerably less time accessing all information relevant to their activity and will still have high confidence that they have identified all relevant material and that no unauthorized changes to documents were made.

- Project managers can use workflows to provide information to technical reviewers and other staff and management, indicating why (whether for review or for information purposes) they are providing the material and due dates for reviewers, and review by their management, as appropriate. The workflows can indicate which reviews can be done in parallel and which need to be performed sequentially.

- Staff members at public meetings and forums can be more responsive to stakeholder questions if they have access to all information in agency systems.

- Seismologists throughout the agency can establish a "community of practice" using a collaborative tool and determine themselves the structure of the site, who should have access to the site, and what types of material the site should contain. Such sites can facilitate regulatory activities and also provide a mechanism to transfer knowledge to and train more junior staff and document this information.

The above scenarios result when staff and stakeholders have seamless access to information and can use the information in workflows they have established in a particular functional area. Having contemporary and historical information available when it is needed gives the staff precedents that can be applied to the current regulatory situation. The application or consideration of precedents helps ensure that agency programs and policies are predictable.

1.3   NRC Policy Framework

The flow of information is critical to the mission of the NRC. In fact, information represents the core mechanism through which the NRC regulates the Nation's civilian use of nuclear materials. Each day, the NRC staff uses information to support licensing and regulatory decisions, to provide evidence in adjudicatory hearings, to document enforcement actions, and to conduct operations. At the most basic level, the effective management and use of information are essential to the NRC being able to meet the safety and security goals and organizational effectiveness and excellence objectives contained in its Strategic Plan.

The NRC's Strategic Plan defines strategies to meet specific safety and security goals and operational objectives. The NRC faces a changing regulatory environment, and to meet these challenges, the Strategic Plan outlines strategies to ensure that the NRC remains a strong, independent, stable, and predictable regulator. This includes having effective and structured regulatory processes in place and ensuring that they are followed. The agency is committed to considering and being responsive to stakeholder input as part of that process. Establishing and maintaining stable and predictable programs and policies that enable stakeholder understanding and participation are key to achieving these goals. Maintaining an open, collaborative working environment also encourages employees to raise safety issues and differing views, strengthening the NRC's ability to regulate effectively.

The NRC's ECM Program will help NRC staff and stakeholders to follow the agency's structured regulatory processes. ECM does this by developing an approach for managing the large volume of content that resides at the NRC in a manner that ensures that this information is available quickly and easily whenever staff or stakeholders need it. The NRC ECM Program vision encompasses content management concepts and practices designed to support agencywide strategies and priorities, including working from anywhere. The program provides a framework for managing information in a way that specifically meets the strategies identified in the NRC Strategic Plan and in NUREG-1908, Volume 1,"Information Technology/Information Management Strategic Plan Fiscal Years 2008–2013," issued April 2008 (the IT/IM Strategic Plan).

The NRC Strategic Plan goals for organizational excellence state that, for participation in the regulatory process to be meaningful, stakeholders need accurate and timely information. Stakeholders must have access to clear and understandable information through active communications and open access to documents and correspondence related to license renewals, license applications, and inspection findings. In addition, operational excellence means that information technology/information management (IT/IM) systems and services must work effectively to deliver information to all participants in the regulatory process. Improvements to information technology and information management at the NRC also offer significant opportunities for increasing the efficiency and effectiveness of NRC operations. As stated in the NRC Strategic Plan, "Timely, high-quality information is critical to the achievement of the NRC's safety and security mission." Without fully supported and reliable content management, the agency will be less efficient in providing the quality, volume, consistency, and timeliness of the information needed to support regulatory processes and activities.

The NRC's ECM Program directly supports the organizational excellence objectives and the openness, operational effectiveness, and excellence strategies contained in the Strategic Plan. The purpose of ECM is to solve business problems, and the program should have a dramatic positive impact on the operations of the various organizations within the NRC and on the staff. In particular, the NRC's ECM Program significantly supports the following strategies described in the Strategic Plan:

- Provide accurate and timely information to the public about the NRC's mission (Openness Strategy 2)

- Provide for fair, timely, and meaningful stakeholder involvement in NRC decisionmaking (Openness Strategy 3)

- Initiate early communication with stakeholders on issues of substantial interest. (Openness Strategy 5)

- Provide clear and timely guidance to applicants and licensees to foster the submittal of high quality and timely applications or license amendment requests. (Effectiveness Strategy 2)

- Cooperate with Federal agencies, States, and Tribal authorities and international counterparts to gain insights and effectively resolve issues to enable the safe and beneficial use of radioactive materials. (Effectiveness Strategy 4)

- Continue to improve the NRC's regulatory and communication programs. (Effectiveness Strategy 7)

- Achieve efficiencies in the licensing process that enable the safe and secure use of nuclear material. (Effectiveness Strategy 8)

- Improve support services to make them more efficient and make it easier to accomplish agency goals. (Operational Excellence Strategy 3)

- Manage agency information and employ technology to improve the productivity, effectiveness, and efficiency of agency programs and enhance the availability and usefulness of information to all users inside and outside the agency. (Operational Excellence Strategy 4)

In addition, other key strategies in the ECM Program will help support the Strategic Plan:

- Provide for one-time entry of information into the ECM.

- Automate records retention, where possible.

- Make available single-search capability, for both staff and stakeholders, for any information in the agency's automated systems.

- Integrate information and business processes across the agency for any one activity, such as new reactor license applications (site reviews; site inspections; vendor inspections; safety, environmental, and security reviews; and associated adjudicatory hearings).

- Integrate information for similar activities, such as data on materials used for spent fuel storage casks or reactor components (information from inspections, operational data on these materials, information reported by vendors and licensees, and technical information that can be obtained through the NRC library and NRC research programs).

The ECM Program also directly supports initiatives outlined by the NRC Chairman. Achieving Chairman Jaczko's goal for the NRC to be a strong regulator depends on improving the agency's financial systems, physical infrastructure, workforce, safety and security culture, regulatory infrastructure, and communication ability. Among his specific objectives are two that are directly related to the ECM Program: upgrading the agency's IT infrastructure, including the current ADAMS, and maintaining a high-performing workforce.

Implementing the NRC ECM Program and replacing the current ADAMS with a modern and fully supported ADAMS application software and operating environment will support the Chairman's goals and objectives. This fully supported operating environment will provide increased reliability and authenticity of agency information accessed by staff, stakeholders, and licensees. It will also provide for more efficient and effective information retrieval by stakeholders wanting to participate in the NRC's technical regulatory processes. From a technology perspective, the new operating environment will support effective, easy to use, maintainable, and highly interoperable agency IT applications. This operating environment will support the management of information throughout its life cycle and empower the staff by providing flexibility in the development of business processes and associated workflows. This new operating environment supports the Chairman's objective to seek opportunities to increase use of new media alternatives and develop the infrastructure to support them in order to allow for more diverse stakeholder outreach and easier access to information.

The Chairman's work-from-anywhere priority will also be supported by the ECM Program. Effectively managing the agency's information through the ECM Program will help enable a business transition from a manual, paper-based basis to the more efficient and effective basis of electronic processing of information. The secure electronic processing of information will help ensure that agency information is stored, managed, and retrieved in a way that is consistent with the needs of a workforce that can work from anywhere. Thus, staff will be able to access information from anywhere. Couple that with the proper technology, and both pieces of what is needed to work from anywhere will be in place.

The NRC ECM Program functionality will also support a high-performing workforce. The program will enhance internal communications; provide business process and workflow tools that will empower staff and should enhance staff engagement; enhance knowledge management initiatives across the agency through the increased use of collaborative tools, virtual workspaces, and better access to information; and enhance opportunities for flexi place and ensure that work-from-anywhere alternatives are available.

The NRC ECM Program also supports IM and IT goals as described in the NRC IT/IM Strategic Plan. Moving to and providing support for a new application software and operating environment to replace the current ADAMS helps meet two strategic goals in the NRC's IT/IM Strategic Plan:

(1)   Information Goal: Make it easy for the staff to produce and access information to perform its work and for stakeholders to participate and interact effectively with the agency.

(2)   IT Applications Goal: Achieve and sustain effective, easy-to-use, and integrated agency IT applications that support the management of information throughout its life cycle.

The information strategic goal to improve ease of access is met by replacing the current ADAMS with a modern system and by providing additional enhanced technologies, such as single search, that improve the ability of staff and stakeholders to access needed information. These improvements will promote a more efficient staff review process and more effective participation by stakeholders in the NRC regulatory process. By modernizing the ADAMS application software and operating environment and automating information management processes, the NRC will have modern IT applications that support information management throughout the information life cycle, as called for in the IT applications strategic goal.

The overall strategy of the ECM Program is consistent with the objective of the NRC's IT/IM program: to manage agency information and employ information technology to improve the productivity, effectiveness, and efficiency of agency programs and enhance the availability and usefulness of information to all users inside and outside the agency. The NRC's ECM vision of providing appropriate and straightforward access to information when needed, regardless of location or access device, is in direct response to all of these enterprisewide business needs and priorities.

2. ECM VISION AND STRATEGIES

The ECM Program vision is the following:

> Improve the quality, efficiency, and security of the NRC's regulatory and administrative processes by providing staff and stakeholders with appropriate and straightforward access to information when needed, regardless of location or access device.

The mission of the ECM Program is to establish a single, trusted framework that will help achieve this vision through enterprisewide guidance, governance, services, and support for the NRC's staff and external stakeholders. The ECM Program is responsible for establishing policies, processes, and activities related to delivering content management solutions in support of the agency's strategic goals and programmatic and administrative activities.

The ECM Program encompasses the different phases of the information life cycle:

- Create: Unstructured, Structured-tables, Web pages, documents, and video

- Store: ADAMS, ADAMS replacement, SharePoint sites, Web sites, databases, and warehouse

- Deliver: Make information available from ADAMS, ADAMS replacement, SharePoint sites, Web sites, and warehouse

- Manage: Compliance with statutes and regulations

- Final Disposition: Transfer to National Archives and Records Administration or destroy

ECM is more than the implementation of one application or system. The NRC's ECM Program is an integrated approach that will combine business strategies, technology tools, projects, policies, governance, integrated records and information management, and staff into a functional framework that will deliver effective and efficient content management.

Modernization and automation efforts that are integral to implementation of ECM and streamlining the ways that staff and stakeholders work within the NRC's regulatory environment include the following:

- automating processes for records management that will then automatically disposition the records based on the retention schedule established in the file plan without the need for manual intervention

- deploying easy-to-use, maintainable, and highly interoperable agency IT applications that effectively manage information throughout its life cycle and integrating all systems in which agency content currently resides

- replacing the current ADAMS, which serves as the hub for a number of systems that support the NRC's core regulatory processes, including licensing, inspections, hearings, and administrative activities, with a new operating environment that enables business process automation and workflow management

- automating processes and workflows to make the agency and individual functional area work processes more efficient and effective

- automating workflows of mission and corporate functional areas, as determined by the lead office in the functional area in coordination with other organizational units working in that area

- redesigning the NRC Web site to improve its usability and to move it from a static information provider to an interactive window into the NRC's regulatory programs and processes (Initially, the redesign will make the current Web site more usable; subsequently, the site will evolve into one that is highly interactive in a way that moves an interested member of the public to become an agency stakeholder.)

- having on-demand availability of information through mechanisms such as single search

- automating e-mail capture as official agency records

- converting hardcopy and microfiche records to electronic versions

- improving collaboration through better document-sharing processes and collaboration software, as well as vastly improved search and retrieval functionality

- increasing compliance with statutes and regulations related to various functional responsibilities such as records management and e-discovery to support adjudicatory licensing hearings

- reducing administrative burdens through improved vendor support, as well as more intuitive user interfaces and improved online help

- increasing operational efficiency through reduced storage costs, improved user account handling, and reduced third-party maintenance costs

3. PILLARS OF ECM

Enterprise Content Management is the ability of an organization to capture, manage, store, preserve, and deliver needed information so that it is available to the right people at the right time. The NRC ECM Program integrates the three fundamental pillars, by bringing together information management, business processes, and technology within a single enterprisewide framework, that will deliver many benefits to the agency and its stakeholders. While the ECM Program provides for integration of these pillars, it will work only if this integration is enacted during the planning, development, implementation, and maintenance of all projects. Section 5 of this NUREG addresses the governance framework that will help ensure successful implementation of the ECM Program.

3.1   Information Management

Information management is guided by the principle that information is a valuable asset that must be secure and made available for efficient and effective decisionmaking. The agency manages information to provide the NRC staff and stakeholders with secure, stable, and predictable access to information for as long as needed. Whether in electronic or physical forms, information must be organized and classified in a reasonable and easily understandable way. This will ensure predictable and consistent search results. There should be consistency in the taxonomy used throughout the agency at a high level and further consistency in specific program areas, such as fuel cycle licensing or new reactor licensing, adopted by all organizational units in the agency that do work in, or related to, these functional areas. Information must also be stored in secure and reliable systems that are created in a way that enables electronic distribution, search, and retrieval across multiple systems.

Information management also addresses how to capture, store, access, and secure information for reuse and distribution. This process can be challenging because of the many forms in which information exists, such as office documents created by individuals, documents created from document imaging with optical character recognition, or information that is created through applications such as electronic Web forms. Information management is also concerned with getting information to the right audience on the right device, such as determining if an e-mail or a portal is the best way to distribute the information. Another component includes managing the tools and techniques used to move content and monitoring those tools' performance.

In addition, information management ensures that the NRC is capturing, maintaining, using, and providing for the disposition (deletion/transfer) of records. This includes the processes for capturing and maintaining evidence of and information about business activities and transactions in the form of records. At the time individuals create or modify information, there should be consideration of whether and how the information should be recorded and captured. The ECM Program must facilitate that process. As an example, when information is generated in SharePoint or when information from ADAMS is modified in SharePoint or any other system, the associated records requirements must be considered.

3.2   Business Processes

Information enables the NRC's program and support offices to conduct their operations and perform their business functions. Under business processes, a key foundation is the workflow capabilities that enable the staff to manage content through its life cycle and also to manage the business processes between the people, systems, and technologies, on the one hand, and the procedures or tasks that drive the flow and management of the information, on the other. These business process technologies must allow for flexibility and include key features such as automating content through business rules and task management within the business process flow. The ECM Program will accommodate the various types of business processes needed throughout the agency. Where possible, the business processes used across the agency should be consistent. In any functional area, and possibly in related functional areas, the business processes used by different organizations need to be integrated. However, some areas will require unique business processes, and it must be recognized that business processes will change, and need to change, over time for a number of reasons.

ADAMS currently serves as the hub for a number of systems that support the NRC's core regulatory and support business processes, including licensing, inspections, hearings, and administrative activities. The agency's ECM Program includes an upgrade to a new application software and a new operating environment, IBM P8, which will enable business process automation and workflow management, improve information access, and streamline application modernization.

The business processes include those specific to operational or mission-critical operations, as well as those needed to manage the information. These processes for managing the information will be embedded and transparent to the user of the operational or mission-critical information. Examples are the NRC business processes that help to facilitate digital signature and authentication of information from various sources, automate records capture and management processes, and enhance collaboration and document sharing. A successful example of ehanced collaboration and document sharing is the virtual workspace created to streamline the review of new reactor applications. If the ECM Program had been in place, this project could have been developed and implemented with far fewer resources and in a much shorter timeframe.

Technology can help facilitate streamlining the management of content, but it is critical that the underlying business needs and strategies must first be identified and analyzed. Very often, these business needs will include a number of associated activities in various organizations within the agency that must be integrated. In 2003, the agency formed a new organizational unit, High-Level Waste Businesss Process Integration, to oversee and coordinate the integration of all agency business activities and IT/IM systems needed for the review of the U.S. Department of Energy's (DOE's) application for a high-level waste (HLW) repository and the associated adjudicatory hearings. It soon became apparent that the HLW program lacked a common understanding of the need to integrate the business activities and processes of the involved organizations; the integrated IT systems and automated business processes that would be needed to support these business activities; and the individual office and agencywide resources required to support this effort. This HLW integration effort required significant resources, particularly for the development of the HLW Meta-System. If a robust ECM Program had been in place, it could have been used to address these needs in a much more effective and efficient manner with a significant reduction in expenditures. The ECM Program Manager will be responsible for monitoring business process needs throughout the agency and, in coordination with organizational units across the agency, assessing appropriate technologies to use for information management and business processes.

3.3 Technology

The technology pillar of the ECM Program provides the systems and tools used to capture, store, secure, manage, and provide access to the NRC's information. Additionally, these systems and tools provide collaboration and knowledge-sharing capabilities to the NRC staff and stakeholders. Under the guidance of the NRC ECM Program, a decision was reached to upgrade the existing ADAMS to the IBM P8 application software to ensure a supportable, scalable, and stable operating environment. The upgraded IBM P8 operating environment is a next-generation, integrated enterprise foundation. It provides the software and hardware needed to implement the agency's business processes and will support the various types of functionality described in this NUREG by using products from IBM or other vendors for areas such as automated records management. Under the ECM Program, technology is designed to establish current, sustainable, transparent, integrated, and secure tools to provide cost-effective user support and automation of business and compliance activities, wherever appropriate. Users need technology solutions with interfaces that are transparent enough to enable them to focus on functionality, rather than on the mechanisms that provide it. A good example is the Web-based interface used for navigating, viewing, and copying new reactor applicant submission documents. This allows the NRC staff to review an application electronically as it was submitted by the applicant. This initiative includes mass processing and support of linked documents submitted via CD. When a Web-based interface was needed for new reactor licensing, under the current ADAMS, significant resources from the NRC's Office of New Reactors and OIS were expended to develop a solution.

Users also need technology solutions that ensure continuity of services in order to minimize disruption and technology solutions for the long-term storage of and access to information. The ECM Program is moving the agency toward adaptable technologies that provide sufficient flexibility to meet changing needs and ensure the long-term sustainability of information, regardless of what other future technologies may be used. Through current and planned investments in ECM-related technologies, the NRC is moving from its current systems portfolio toward a system that significantly improves effectiveness and efficiency in managing information. While the new system will provide more flexibility than in today's ADAMS, it must be recognized that any system has finite flexibility and that the agency must avoid the proliferation of custom code that has occurred since ADAMS was deployed. The proliferation of custom code in today's ADAMS has significantly contributed to the problems in supporting and maintaining a stable system. The use of custom code in P8 would make the P8 operating environment much more complicated and costly to maintain and upgrade throughout the planned ECM lifecycle.

The ECM Program will be implemented through a phased rollout with increasing scope and capabilities. The initial phase of this multiyear strategy includes mirror and operate P8 Library and ADAMS Main Library concurrently; transition existing applications to P8 Library (the Document Processing Center, Publicly Available Records System, Electronic Hearing Docket, e-mail capture, and others); transition existing operating procedures and processes to P8 Library; train and transition users to P8; and decommission the ADAMS Main Library. This initial phase will also implement enhanced browser-based document and package management features and reduced sign-on.

3.4   Core Components of ECM

An ECM Core Component Model has been developed for the implementation of the NRC's ECM Program. The model integrates the "three pillars of ECM" discussed previously. The core components are information management, which has three subcomponents (document and image capture, document and image management, and information); records management; electronic forms management; Web content management; portals/collaboration; search; governance; and classification/taxonomy/metadata. For each component and subcomponent, the model addresses strategy, business area, processes, systems, tools, owners, stakeholders, and ECM service offerings. This model will facilitate not only an understanding of the core components of ECM, but also a much better understanding of which business areas, processes, and ECM service offerings each of the core components supports, as well as the systems that support this functionality. It will give staff and management insights into how they can best use the NRC's ECM Program. In addition, it clarifies which organizations and systems need to support various components of the ECM Program.

4. ECM PROGRAM STRATEGIES

The NRC ECM Program, in pursuit of goals that address explicit business needs, will implement five strategies for success. First, by enabling the shift from paper-based, manual business processes to more effective, efficient, and user-friendly electronic processing of information, the ECM Program intends to **transform** the NRC and the way it does business. The ECM Program also intends to **enable** internal and external stakeholders to successfully handle and process the significant growth in information from increased business activity. Third, the ECM Program intends to ensure transparent and timely **information access** necessary for the agency to carry out its mission, regardless of location or format. In addition, the ECM Program will facilitate **information sharing and collaboration** among staff and stakeholders during daily operations by providing appropriate technology tools and accurate information about, and support for, those tools. Finally, the ECM Program will ensure a **robust application infrastructure** to support evolving technology needs consistently and uniformly across the agency. Several initiatives are currently underway in support of these strategies, as described next.

The NRC's ECM Program comprises different projects, activities, and services that support sound management of the agency's content. Some will be introduced as part of the transition to the IBM P8 operating environment; the timing of others will depend on future budgets. The following describes implementation of key activities that support the ECM Program.

Operational Maintenance

Maintenance of ADAMS has as its objective, replacement of the current application software with a supported application software, IBM P8, before the current application software becomes unsupported. The FileNet software that supports the current ADAMS will become unsupported by December 2011.

Besides the move to a supported application software, activities in the maintenance project include (1) the migration of documents and files stored in the current ADAMS to the supported IBM P8 and (2) migration of the 17 applications that rely on the current ADAMS to IBM P8. During migration, a synchronized library will be in place for both current ADAMS and IBM P8 to support those applications not migrated that still rely on the current ADAMS.

Modernization

A key objective of the modernization project related to ECM is to automate workflows, which will make the agency and individual function work processes more efficient. The project involves automating processes and workflows. Examples of modernization activities include document profiling and capture of e-mails as official agency records. In addition, workflows of mission and corporate functions will be automated as identified by lead offices. There will also be a business process manager component that provides the capabilities to not only automate, but also monitor and report on the business processes that the NRC uses to pass content from one participant to another for actions according to a set of rules. Repeatable workflows present an opportunity to automate manual tasks and activities, thereby increasing staff efficiency and effectiveness. The advantages of automating more routine manual workflows include the following:

- Support planning activities for better workload management.

- Ensure that the information goes to the proper organizations and individuals for a stated purpose, such as for information, review, approval, or other action.

- Establish a standard level of service expectation based on the workflow execution.

- Provide information that can be used to continuously improve the workflow process to make it more efficient.

- Provide an automated means to measure and quantify the amount of work being processed and the resources required to do the work.

- Increase the consistency of the work products and results produced.

The modernization project will also usher in a reduced sign-on concept. This feature will be available when the move to the new IBM P8 operating environment is complete. As part of the transition to modern technologies, the program will use reduced sign-on for all applications that are part of the ECM Program. This will result in fewer separate login accounts that staff need to manage, as well as an improved security posture for the agency. Information access will be strictly managed so that only users who should have access to certain information will have it. As business activity increases in the coming years, the new system replacing the current ADAMS will increase functionality and flexibility to handle this growth.

Automated Records Processes

Modern enterprise content technologies have the capability to automatically capture agency records and apply classification and retention rules so that they are retained, destroyed, or transferred based on business needs and federally mandated records management requirements. This automated capture and profiling will benefit the NRC staff by reducing the need to make record status and profiling decisions, promote compliance with Federal records statutes and regulations, reduce the amount of information no longer needed that becomes a disclosure risk, and minimize the perception that records management compliance applies only to information and content at the end of its life or only to "official" agency records.

Activities in this area involve the ongoing revision, or if needed, development of file plans which establish the records retention schedules. In addition, automated document profiling will ensure the placement of documents in the appropriate retention schedule. The electronic records processes will then automatically disposition the records based on the retention schedule established in the file plan without the need for manual intervention.

As the NRC progresses to using more and more systems for collaboration or communications, the ECM Program will seek ways to apply records management methodologies (such as retention rules) to this information. The NRC's Records Management Program is developing draft recordkeeping guidance for collaborative tools such as wikis, blogs, and community-of-practice portals. Rules will be embedded into systems and tools that will apply guidance and policies without requiring workarounds or staff decisionmaking about record status.

New mechanisms, such as automated capture of e-mail or documents in applications such as Microsoft Word and SharePoint and on other sites, will ease the burden on users and streamline document management processes throughout the agency. Improved document profiling, better version handling, and automated record capture will increase the reliability and authenticity of agency information, while increasing compliance with responsibilities such as responding to discovery and disclosure requests, as well as indentifying records eligible for destruction or transfer to the National Archives.

Digitization

Digitization is an ongoing service. The objective of digitization is to move hardcopy documents into an electronic format so information is promptly available to staff who can then work from anywhere and to NRC stakeholders. Documents are digitized as they are retrieved from storage and as needed, or as requested by offices based on needs. Examples of office requests are the digitization of all hardcopy files for the Watts Bar and Indian Point nuclear power plants. Digitizing key collections will also help to enhance the efficiency of the regulatory process by allowing more than one person to access documents concurrently. Digital content also allows efficiencies of organization, classification, storage, search, retrieval, retention, and disposition of agency records. The ECM Program is working with the program offices to identify priority documents to be digitized. Older NUREG-series technical reports prepared by the NRC staff and the National Laboratories (Argonne, Brookhaven, Idaho, Los Alamos, Oak Ridge, Sandia, and others) are also a priority because of their high volume of use. By digitizing the paper copy for official recordkeeping, the agency will be confident that the most accurate, complete, and trustworthy content is accessed.

Web Redesign

The objective of the NRC Web redesign effort is to improve the usability of the Web site and to move the NRC Web, a static information provider, to an interactive window into NRC regulatory programs and processes. Initially, the redesign will involve revising the current Web site to make it more usable. Subsequently, the site will evolve into one that is highly interactive in a way that encourages an interested member of the public to become an agency stakeholder.

Network Drive Integration

The objective of network drive integration is to store information in a single repository to eliminate the need for multiple servers to store information. A second objective is to provide for efficiency in staff retrieval of information for FOIA requests, hearing discovery, and general information needed to conduct the NRC's work. Network drive integration is exploring how to more efficiently handle content now stored on individual network drives.

Collaboration Tools

The NRC has launched projects for Web-based collaboration, using Microsoft SharePoint and Tomoye, and developed a SharePoint site for management, coordination, and governance of NRC SharePoint sites. SharePoint offers a platform for sharing information and working in teams through collaborative workspaces, portals, and social computing tools and is integrated with the full Microsoft Office Suite used throughout the NRC. For example, OCFO is using SharePoint's collaborative features to obtain information on the current and future budget calls. The NRC staff is also using Tomoye as a community-of-practice platform (known as the NRC Knowledge Center) in support of agencywide knowledge management initiatives. Together, these two applications offer a wide range of collaborative functionality that is only beginning to gain momentum. With the deployment of SharePoint and Tomoye, NRC staff members will have the software needed to easily share agency documents while they are working on drafts or if the material is frequently referenced. These tools will also give staff members the means to work together in communities of practice, sharing ideas, transferring knowledge, training new staff, and fostering knowledge management for years to come. In addition, the use of these tools can minimize the amount of information kept on shared drives, in the e-mail system, and in other disparate systems.

Collaboration tools, when integrated with business and records retention rules, will allow for control and life-cycle management of the information regardless of where the content resides. Integration of these tools with (1) the current ADAMS replacement for information storage and (2) the records program are two significant steps in ensuring that the information generated by these tools is properly stored and managed, where appropriate, as records. The objective of this integration is to place information in a single repository to ensure that it is properly stored, managed, and dispositioned in accordance with the agency's records program. OIS has already developed a SharePoint Program. Its mission is to establish a single, trusted resource for all SharePoint guidance, services, and support. The SharePoint Program will deliver a specific set of capabilities related to content management, collaboration, and communications across the agency.

Hardcopy Record Storage

Hardcopy record storage ensures that these agency records are stored, managed, retrieved, and dispositioned consistent with agency needs and Federal requirements. The objective is to make this information available to staff and stakeholders when they need it. This is an ongoing activity.

Single-Search Capability

The NRC staff needs the ability to simultaneously search multiple online databases or Web resources and other document collections or repositories. Single-search capability is expected to give staff and stakeholders the ability to find and retrieve all information stored in various repositories by using one search. Work on the single-search capability includes the development and deployment of the single-search capability both internally (for staff) and publicly (for agency stakeholders). This capability will allow the user to specify searches by data contained in the content profiles (the metadata describing the information in the content) and in the content itself. The staff content search tool will be engineered for knowledge workers. Under the ECM Program, modern enhanced search technologies will be developed to permit a single-search functionality that will ensure search results that are predictable, consistent, and repeatable. These modern search technologies may identify relevant and important information of which the individual doing the search was unaware. Development of a taxonomy and data standards and implementation of automated records classification will greatly enhance the retrieval of relevant information.

Taxonomy

Taxonomy is a categorization framework that organizes content. An enterprise taxonomy structures an organization's content for use throughout the document life cycle, including document creation, collaboration, security, access, publishing/storage, retrieval, and disposition. The objective is to develop a high-level agencywide taxonomy, but allow program and support offices to organize content to suit their functional needs. In any functional area, a lead office would develop the taxonomy in consultation with other offices working in that functional area, or in some cases, related functional areas. OIS has launched an effort in this area and will coordinate it with other offices.

Work from Anywhere

The ECM Program will also enable the agency staff to securely access and use the new technology and systems that will improve work performance regardless of where they are physically located. In other words, the idea is to provide a "virtual workplace" capability to individual users. Currently, the agency's vision is to begin to roll out Web-based applications which are much easier to integrate in remote access technologies, e.g., Citrix, which improve employee mobility. Digitization also supports this vision so that information is available on line and not stored on paper at a warehouse. While deployment of several supporting technologies should happen in the near term, the overall vision will take several fiscal years to realize.

Once implemented, the projects, activities, and services described above will result in an ECM Program that is responsive to the needs of the agency, its staff, and its stakeholders. This ECM Program should satisfy the program drivers that were developed by OIS, in coordination with all other agency organizations, when it became apparent that the current ADAMS needed to be replaced.

Information on the current status of NRC projects and activities related to ECM capabilities and technologies will be posted on the NRC Intranet. This information will be updated to reflect the current status of key ECM activities and will include an up-to-date graphic ECM deployment schedule that will show when the various additional functionality discussed in this NUREG will be available to staff. Deployment schedules will depend on availability of budgeted resources.The NRC Public Web site will be updated in a timely manner to keep stakeholders informed of enhanced or new functionality relevant to them.

5. GOVERNANCE

OIS has responsibility for management of the ECM Program in support of the agency's mission, objectives, and goals. To achieve and implement the vision for ECM, OIS is establishing a governance framework that will provide overall direction and oversight for all ECM-related activities, from planning through implementation and maintenance. It will integrate the three fundamental pillars of content management by bringing together information management, business processes, and technology within a single enterprisewide framework, ECM. This includes developing the framework for the ECM Program; establishing policies, processes, and procedures for the program; and providing oversight, coordination, and assistance for projects and activities related to delivering content management solutions in support of the agency's programmatic and administrative support activities. All initiatives will be performed in an open manner in coordination with the other offices and the regions. This governance framework will be integrated with, or into, the overall IT/IM governance framework.

To realize the benefits of an ECM Program, staff and management throughout the agency will need to operate in accordance with this governance framework. While there are firm requirements, such as those related to IT and IM security, including access contols, and the need to manage information throughout its life cycle, there will also be more flexible and general requirements related to development of business processes and the related workflows. Staff and management in the program and administrative support offices and regions, who are the experts in their business areas, will have the flexibility to determine what business processes and workflows, including collaborative sites such as SharePoint, are needed to conduct work activities in their functional areas. In each major functional area, such as reactor license renewals, a lead office should coordinate development of file plans and taxonomies for that functional area. There should be a high-level taxonomy established agencywide and then an additional layer(s) of taxonomy for each major functional area. The ECM Program Manager will need to ensure that consistent taxonomies, as appropriate, are used in content management and basic content service deployments and extend content governance and best practices across the enterprise. Staff and management throughout the agency must all guard against software purchases or development activities that isolate information or user populations.

In addition, the ECM Program Manager will work with both internal and external stakeholders to promote the program, resolve program- and stakeholder-related issues, ensure that stakeholder needs and requirements are fully identified and addressed, manage ECM communications and outreach, coordinate user training and support, and employ change management tools to increase user acceptance and improve services.

6. CONCLUSION

Current trends in applications for new nuclear power plants are keeping the NRC focused on the future. The workload has increased and public and congressional scrutiny has heightened as the result of the receipt of 18 combined license applications for 28 new reactors, renewal applications for existing power plants, applications for new fuel cycle facilities, and an application from DOE for an HLW repository. (Of the 18 applications for combined licenses, five applicants have requested that the NRC suspend its review of their applications given changing business strategies.)The agency has also instituted a robust knowledge management program to increase knowledge capture and retention in view of the NRC's rapidly changing workforce. The NRC has an increasingly mobile workforce in need of flexible tools for collaboration without boundaries.

Outdated technologies and the increased use of Web-based applications have also led the NRC to reexamine how it does business, with an eye toward improving business processes and information management. Significant challenges for the agency in the next 5 years require prudent planning to ensure efficient and effective use of all agency resources, including existing and future information systems and technologies. As these challenges gather momentum, the agency needs to leverage new technologies, while moving forward with a sound business framework.

The NRC's ECM Program provides the essential framework and activities that will help keep the agency focused on its mission and able to achieve its goals.

7. REFERENCES

1.   U.S. Nuclear Regulatory Commission. NUREG-1614, Volume 4, "Strategic Plan: Fiscal Years 2008–2013." Washington, DC. February 2008.

2.   U.S.Nuclear Regulatory Commission. NUREG-1908, Volume 1, "Information Technology/Information Management Strategic Plan Fiscal Years 2008–2013." Washington, DC. April 2008.

NRC FORM 335
(9-2004)
NRCMD 3.7

U.S. NUCLEAR REGULATORY COMMISSION

BIBLIOGRAPHIC DATA SHEET

(See instructions on the reverse)

1. REPORT NUMBER (Assigned by NRC, Add Vol., Supp., Rev., and Addendum Numbers, if any.)
NUREG-1954

2. TITLE AND SUBTITLE

Title: Enterprise Content Management
Subtitle: Streamlining How Staff and Stakeholders Work within the Nuclear Regulatory Commission's Regulatory Environment

3. DATE REPORT PUBLISHED

MONTH	YEAR
September	2010

4. FIN OR GRANT NUMBER

NA

5. AUTHOR(S)

John Linehan and Information and Records Services Division Staff

6. TYPE OF REPORT

Information Management

7. PERIOD COVERED *(Inclusive Dates)*

2010-2015

8. PERFORMING ORGANIZATION - NAME AND ADDRESS *(If NRC, provide Division, Office or Region, U.S. Nuclear Regulatory Commission, and mailing address; if contractor, provide name and mailing address.)*

Information and Records Services Division
Office of Information Services
U.S. Nuclear Regulatory Commission
Washington, DC 20555-0001

9. SPONSORING ORGANIZATION - NAME AND ADDRESS *(If NRC, type "Same as above"; if contractor, provide NRC Division, Office or Region, U.S. Nuclear Regulatory Commission, and mailing address.)*

"Same as above"

10. SUPPLEMENTARY NOTES
NA

11. ABSTRACT *(200 words or less)*

This report provides staff and stakeholders with an understanding of Enterprise Content Management (ECM), describes the U.S. Nuclear Regulatory Commission's (NRC's) planned ECM Program, and discusses how this ECM Program will provide critical support to the NRC's mission and streamline the work of staff and stakeholders within the NRC's regulatory environment. ECM is the ability of an organization to capture, manage, store, preserve, and deliver needed information so that it is available to the right people at the right time.

12. KEY WORDS/DESCRIPTORS *(List words or phrases that will assist researchers in locating the report.)*

ECM, Enterprise Content Management, Information Management, IM, Content Management, CM

13. AVAILABILITY STATEMENT

unlimited

14. SECURITY CLASSIFICATION

(This Page)

unclassified

(This Report)

unclassified

15. NUMBER OF PAGES

16. PRICE

NRC FORM 335 (9-2004)

PRINTED ON RECYCLED PAPER